Irene Smets

AF955905

The Royal Greenhouses, Laeken

Ludion

Preface

The Royal Greenhouses in Laeken are amongst Belgium's most important 19th-century monuments. Their construction entirely in metal and glass was spectacularly innovative for its time. Although the technique was used elsewhere in Europe, the scale and sheer architectural quality of the Greenhouses have rendered them unique.

The rise of botany as a science in the 18th century and intensive European colonial activity in the 19th sparked a lively interest in exotic plants in the West. There was a veritable craze among the aristocracy for cultivating orange trees, which were kept in orangeries during the winter. The Orangery at Laeken dates from the early 19th century, but it was also used to house the orange trees that Charles of Lorraine imported in the 18th century during his time as the region's governor. Tropical and above all subtropical plants became fashionable a little later. Although these needed extra light, heat and space, modern iron and glass structures meant that they were able to survive in the Northern European climate.

King Leopold II, who commissioned the building of the greenhouse complex, had a special love of plants and flowers. He took an active interest in the choice of plants in the Greenhouses and also in the gardens and grounds of his royal residences. We know from a variety of sources that he inherited this fascination from his parents.

The King was especially fond of his camellia collection. He was known to rise in the middle of the night to check whether they were suffering from the cold, and he regularly brought new varieties back with him from his foreign visits. By the time of his death, the collection had grown to include a thousand different types.

The damp climate and the humid atmosphere inside the Greenhouses mean that they have always required a great deal of maintenance. The Belgian Buildings Authority has been working intensively for several years to restore the large greenhouses, which have been brought back to their magnificent, original condition. The modern techniques used to achieve this are a tribute to the know-how of Belgian companies.

Leopold II was keen to share the joy that the world of plants brought him. To this end, he had several reception rooms built near the Greenhouses, so that he could receive a large number of guests there. The King died, however, before the construction work was complete. The present King and Queen also consider this to be the perfect location for celebrations, such as the reception to mark the King's 65th birthday and the wedding reception of Prince Philippe and Princess Mathilde in 1999.

The Greenhouses have been opened to the public at the express wish of the royal couple who, like their predecessors, are keen to share with you their love of flowers and plants.

J.F. Willems
Grand Marshal of the Court

p. 2 The gallery of the Winter Garden.

The dome of the Winter Garden.

p. 6 *Medinilla magnifica* in the Embarcadère.

Contents

The chateau at Laeken is first glimpsed as you pass through the main gates into the estate. The Royal Stables are located to the right – largely concealed by the trees. The Royal Greenhouse complex adjoins the north-east wing of the chateau. The rural-style studio of Queen Elisabeth and the Palm Pavilion, in which King Leopold II spent the final years of his life, form part of this grandiose structure of metal and glass.

The estate is much larger than is apparent from this vantage point. During your visit to the Greenhouses, you will discover another part of it and will be treated to some lovely views.

The chateau was originally known as 'Schonenberg'. It was built in the 1780s as a summer residence for the Austrian governess, Maria-Christina – daughter of Empress Maria-Theresa – and her consort, Duke Albrecht of Saxony-Teschen. The Duke wrote in his memoirs that he and Maria-Christina had bought some land 'in the most beautiful and well-favoured location in the [Brussels] area, we had a house built and a beautiful garden laid out in the English style'. The celebrated English gardener,

King Albert II and Queen Paola

Lancelot 'Capability' Brown, designed the grounds.

The building was abandoned following France's annexation of the Austrian Netherlands, and subsequently fell into disrepair. In 1803, however, Napoleon Bonaparte gave instructions for the chateau to be bought and refurbished. Following the defeat of Napoleon in 1815, the building, which now belonged to the government, became the summer residence of King Willem I of the Netherlands, who had a large orangery built a hundred metres or so from the chateau.

Belgium gained its independence in 1830 and Laeken subsequently became a residence for the new country's sovereign. The first King of the Belgians, Leopold I – who ruled from 1831 to 1865 – did little to the chateau itself, although he did purchase a number of adjacent properties. His son Leopold II (who reigned from 1865 to 1909) had the chateau enlarged and refurbished on a number of occasions, not least because of the damage it suffered during a fire on 1 January 1890. He also extended the estate, had a huge greenhouse complex constructed and had the estate remodelled by a French garden designer. The final large-scale building programme began in 1902, when the two wings were added, incorporating a number of new features, such as the Royal Stables. Leopold II intended to use Laeken as a kind of Palace of the Nation, which would host conferences and important ceremonies. To that end, he considered having a railway built, which would lead to an underground station near the Grand Staircase of the chateau's Long Gallery. The project was, however, abandoned on the King's death.

At the end of his life, Leopold II presented his royal possessions to the State, in what became known as the 'Royal Donation'. The foundation of the same name still manages a substantial proportion of the Laeken estate.

The latter is 220 hectares in size – equal to the entire surface area of Monaco – and takes in the chateau and the associated buildings (Prince Philippe and Princess Mathilde live in one of the wings), the gardens with the Greenhouses, the Belvédère, in which King Albert II and Queen Paola live, and Stuivenberg, which is the home of Queen Fabiola and where Princess Astrid and her family will eventually live.

Detail of the wrought ironwork in the Winter Garden.

The esplanade in front of the Embarcadère

Until the beginning of the 20th century, members of the Court travelled around by coach or horse. Care of the royal horses and coaches was the job of the Chief Equerry.

In 1902, Leopold II commissioned the architect Charles Girault to carry out the large-scale extension and refurbishment of the Chateau at Laeken. New and more spacious stables were one of the items on the King's shopping list. Girault incorporated the new complex in the south-west wing of the castle. The buildings were nowhere near completion when Leopold II died in 1909 and so work continued under Albert I. Charles Girault withdrew from the project and the building was completed by Octave Flanneau. The Royal Stables comprised a riding school, stables, a tack room, a smithy, a saddle room and, on the first floor, rooms for the Chief Equerry's staff.

The era of the horse and coach was, however, coming to an end. The early 20th century saw the rise

of the automobile, and the riding school and part of the stables were soon given a new use. The Chief Equerry's department was abolished in 1951 and management of the garages became the responsibility of the Commandant of the Palaces.

The two large horse-boxes at the back bear the names of two horses that were presented to King Baudouin in 1975 by King Khaled of Saudi Arabia: Annaz and Gamilla. The pair produced the foals Sahel, Sarepta, Leila, Kalyan and Dalika, all of whose names – apart from Dalika's – appear on the nameplates. The horses were stabled in the winter and spent the summer in a field. They were regularly saddled up for rides around the estate. Godailleur, Cleansteel and As de Trèfle were ridden by members of the Royal Family in the 1950s. Mickey, Bella and Dahab lived in the stables between the two World Wars. Mickey and Bella were ponies and were generally used to pull carriages with children. Dahab was a saddle horse presented as a gift from Egypt.

The stables also contain several 19th-century state coaches, including the Gala Berline 'Louise-Marie', which carried Queen Astrid to her wedding at the Church of St Gudule on 4 November 1926.

Other rooms in the stables contain interesting objects and costumes from the former Chief Equerry's department.

The tack room.

Orangery

The next point of interest after the stables is the Orangery. The esplanade in front of the building is lined with orange trees, bay trees, rhododendrons and camellias. Many of these beautifully maintained ornamental plants are rare antiquities. Some of them are brought into the open air in mid-April, with the rest following in mid-May. They stay outside until the middle of October, before spending the winter at a constant 3 ° C in the Orangery and two outbuildings.

The elongated façade of the Orangery is broken up by 26 pilasters. The two 'minarets' you can see rising above the building are the iron chimneys of the Greenhouses' heating system.

The collection of orange trees consists primarily of two types – orange *(citrus aurantium)* and mandarin *(citrus reticulata).*

The camellia *(camellia japonica)* is an evergreen plant with glossy, leathery leaves and striking flowers, the colour of which varies between white and dark red. There are types with single, half-double and double flowers. The plant originated in China and Japan, but many cultivars (cultivated varieties) were developed in Europe, including Belgium. Laeken's camellias flower from December to May.

Inside the Orangery, between the orange and bay trees, and the fragrant rhododendrons *(rhododendron fragrantissimum),* oleanders and camellias, there is a bust of Leopold II by the sculptor Jef Lambeaux.

The Theatre Greenhouse, the winter home for some of the camellias (not open to the public), is located at the east end of this large space.

The Banquet Greenhouse.

The Banquet Greenhouse is located at the west end. It is used from time to time for celebrations and receptions, such as the one held in 1999 to mark the wedding of Prince Philippe and Princess Mathilde.

Passing through the Tree Fern Annex leads you to the Winter Garden. At this point, you simply pass through the gallery that runs around this greenhouse – you will have the chance to explore the whole Winter Garden at your leisure at the end of your visit.

The façade of the Orangery with the two iron chimneys of the large greenhouses at the top. There is a beautiful hothouse azalea in the foreground.

Camellia japonica.

Winter Garden

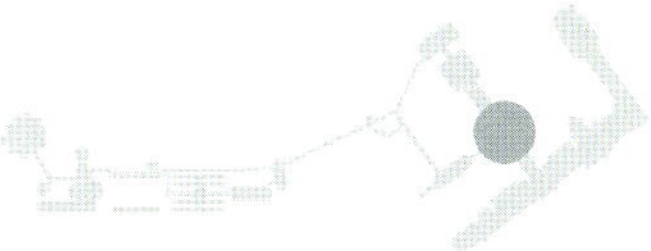

'The greenhouses at Laeken are the loveliest monument ever erected in honour of the study of exotic plants', a famous botanist wrote in 1883.

The Winter Garden or Grand Rotunda was designed by the architect Alphonse Balat for Leopold II. It was built in 1874–76, making it the first greenhouse in the stunning glass city that was to grow up on the Laeken estate over a total of 30 years. The year '1876' appears in the central mosaic with Leopold II's monogram. The structure was intended from the outset as a site for royal receptions.

A huge metal framework comprising 36 beams supports the glass dome. The beautifully decorated iron construction is visible from both the inside and the outside of the rotunda, with flying buttresses running across the gallery. A royal crown can be seen on the lantern of the dome. Inside the greenhouse, the weight of the dome is supported by a circle of stone pillars with Doric capitals. These robust columns lend stability and strength to the iron and glass architecture. Fine iron and glass arcades stand on the entablature of this colonnade, forming the transition between the stone pillars and the structure of the dome.

You walk through an unreal space – evocative at once of the past and present, filled with the light that enters from outside, and per-

meated with the scent and humidity of dense tropical overgrowth under glass.

The scale of this greenhouse is sufficient to allow tall palm trees to grow. Most of the specimens date from the time of Leopold II. Near the passage leading into the Orangery is a large Canary Island date palm *(phoenix canariensis)* – a powerful, slender tree with a crown of upright, pinnate leaves. The fan palm (*livistona chinensis,* from Japan and China; *livistona australis* from Australia) has folded and deeply laciniate leaves two metres long, with hanging fronds at their tips. The Kentia palm (*howea forsterana* and *howea belmoreana,* from Australia) has characteristic rings around its trunk and pointed, pinnate leaves. Another interesting tree is the ladyfinger (*musa paradisiaca,* from Asia; wide, very long leaves). Near the passage into the Congo Greenhouse, partially hidden behind the palm trees, you can see a fourteen-metre-high cinnamon tree (*cinnamomum zeylanicum,* from Sri Lanka; the young leaves are pink and green). The gallery opposite contains the extremely rare aralia (*oreopanax dactylifolius,* originally from Mexico, difficult to cultivate; digitate leaves with terminal umbels). Various types of fern grow at the foot of the majestic trees.

The soil of the gallery is covered with a marvellous carpet of flowers. Beginning in January, yellow, white and pink varieties of cymbidium flower among the rocky walls. You will also find datura (South America; funnel-shaped yellow flowers), hedychium (Himalayas; dozens of lemon-yellow flowers with bright-red filaments, on a long stem), the bird of paradise plant (Canary Islands; yellow-orange flowers), the butia (*cocos bonnetii,* from Brazil; palm tree with a short, thick trunk, full crown with decoratively curved leaves) and a very old oak fern with orange spurs on the underside of its deeply laciniate blue-green leaves *(polypodium glauca)*.

The large Palm Tree Annex, which mainly houses tree ferns and royal ferns, links the Winter Garden with the Congo Greenhouse. Leopold II's monogram, flanked by lions, appears in wrought iron above the entrance to this linking greenhouse. The same motif is worked into the arch that leads into the Tree Fern Annex, linking the Winter Garden and Orangery.

The young leafstalks of the royal fern (*cibotium regale,* from Mexico and Guatemala) uncoil attractively. Their reddish hairy skin protects them from rodents. In the wild, this plant is threatened with extinction, but the gardeners at Laeken succeeded several years ago in cultivating young plants from spores – a rare achievement.

The Winter Garden or Grand Rotunda, from A. Vierendeel, *La construction architecturale en fonte, fer et acier*, pl. 56, Paris-Brussels, 1897–1902.

Apart from the Orangery, which was built around 1818 for Willem I of the Netherlands, and a number of small hothouses from the time of Albert I (built just before the First World War), the entire greenhouse complex was built between 1874 and 1905 on the initiative of Leopold II. The King followed the construction and planting of the Greenhouses with intense interest for fully thirty years, sparing neither expense nor effort to make his glass city something unique.

c. 1818	Construction of an orangery and theatre in the classical style by the 'Architect of the Royal Palaces', Ghislain Joseph Henry from Dinant. Willem I commissions the orangery – one of the largest in Europe

1868	Leopold II decides to build a Winter Garden at Laeken and selects Alphonse Balat, 'Architect to the King', to head the project. The latter is not ready for commencement for several years
1874–76	Construction of the Winter Garden (A. Balat)
1876–80	Completion, interior work and planting of the Winter Garden under the English garden designer John Wills
from *c.* 1878	Alteration of the Orangery and Theatre; construction of the Camellia Greenhouse, the Banquet Greenhouse and the Theatre Greenhouse; the Orangery is linked to the Winter Garden by the Tree Fern Annex (A. Balat)
20 May 1880	Ceremonial opening of the Greenhouses; Reception to mark the betrothal of Princess Stéphanie and the Austrian crown prince, Archduke Rudolf, in the Winter Garden
1885	Construction of four hothouses, including what is now the Azalea Greenhouse (supplied by the Compagnie continentale d'horticulture from Ghent)
1886–87	Construction of the Congo Greenhouse, the Palm Tree Annex, which links it to the Winter Garden, and the Embarcadère (A. Balat)
24 Dec. 1891	Second opening of the Greenhouses in the presence of the patrons and the workers who built the complex
1892	Alteration of the entrance to the Orangery; the royal monogram above the round arch dates from this time (A. Balat)
from 1892	Leopold II organises regular receptions in the Greenhouses and opens them on a limited basis to the public
1892–95	Construction of the Palm Greenhouse, Débarcadère, Diana Greenhouse, Rhododendron Greenhouse, several hothouses and the galleries between the Palm Plateau and the Winter Garden complex
1895	Completion of the Iron Church – the final achievement of A. Balat, who died that same year
1902	Construction of the Maquet Greenhouse (a camellia nursery adjoining the Winter Garden) and the adjacent New Orangery, designed by Henri Maquet, 'Architect-Consultant to the King'
1905	The architect Charles Girault replaces Balat's Theatre Greenhouse with a new one – the current Theatre Greenhouse
1909	Leopold II dies on 17 December in the Palm Pavilion, in which he lived for the final years of his life

By the time of Leopold II's death, the Royal Greenhouses in Laeken were the largest and most remarkable private botanical garden in the world. Apart from the Camellia Greenhouse, the Iron Church and a few small hothouses, the greenhouse complex has remained almost entirely intact to this day.

The Congo Greenhouse is a square space surmounted with an octagonal dome, surrounded by four small, square domes. The lantern of the central dome is decorated with a star – symbol of the Congo Free State. You will also find the star inside the structure, in the round iron arches that support the dome structures. In each case, it is used as a kind of keystone.

Alphonse Balat built the Greenhouse in 1886–87 to provide the King with a space in which to place flora from the Congo Basin. However, the tropical trees and plants from that region failed to flourish in the Greenhouses, as they need a lot of light and heat in the winter, too. For that reason, the Greenhouse was filled instead with subtropical vegetation. It currently houses large Australian palm trees, rubber plants and ferns, on a bed of spikemoss *(selaginella kraussiana)*, interspersed with annuals.

The central path of the Greenhouse runs straight through the Palm Tree Annex to the Winter Garden. In the opposite direction, it takes you to a wide staircase, luxuriantly bordered with flowers. The Embarcadère Greenhouse is located at the bottom of the white stone staircase.

Leopold II was fascinated by the unexplored regions of Central Africa. Motivated by an interest in geography, a desire to spread Western civilisation, and economic and political ambitions in the spirit of the time (the acquisition of a colony for Belgium), around 1876 he began to work with geographers and explorers – the best known being Henry Morton Stanley – on the exploration and opening up of Central Africa. The Congo Free State, with Leopold as its sovereign, was recognised at the Conference of Berlin in 1885.

The King was mesmerised by the mystery of equatorial Africa, with its exotic fauna and flora and its inexhaustible wealth – ivory, copper, rubber, tropical hardwoods, and so on. It is against this background that we have to understand the construction of the Congo Greenhouse. The Congo was not transformed from a personal royal possession into a Belgian colony until 1908, remaining so until 1960.

Flowers were his poetry and his antidote, as it were, to the demands of a harsh reality.
(H. Carton de Wiart)

King Leopold II (9 April 1835–17 December 1909; King of the Belgians from 1865 to 1909) was very fond of exotic flowers and plants. He was proud of the large collection he assembled in his greenhouses. 'The collection of orange trees at Laeken,' he wrote in 1908, 'like the camellia collection, is the loveliest in the continent of Europe. The orange trees are incomparably more attractive than those at Versailles or the Tuileries, while the camellias are wholly unique. No other such collection is to be found, even on the Lago Maggiore.'

Most of the orange trees had been at Laeken since the 18th century. Willem I of the Netherlands had a new orangery built for them around 1818. Leopold II extended the collection.

The camellia collection was largely Leopold II's own work. By the end of his life, there were about a thousand camellia plants in the Greenhouses.

When the planting of the Winter Garden began in 1879, the King called on the leading botanists, nurserymen and suppliers. Many of the plants were donated by botanical gardens in Belgium and abroad. The Kruidtuin in Leuven and the then famous greenhouses of the chateau of the Duke of Arenberg in Enghien supplied a number of large palm trees. Many of the palms that you see today in the Winter Garden are either the original trees or descendants of the first ones.

The orange trees, camellias, palm trees and tree ferns have been the glory of the collection from the beginning, and they remain the most important ensembles at the Royal Greenhouses. There are, however, countless other rare and extremely beautiful subtropical and tropical trees and plants here as well.

Cymbidium orchid.

Congo Greenhouse.

Embarcadère

The Embarcadère Greenhouse dates from the same period as the Congo Greenhouse. When receptions were held at the Winter Garden or the Banquet Greenhouse, it was here that the King's guests entered the complex. Part of the greenhouse was used at that time as a cloakroom.

The structure consists of two compartments. The innermost compartment has a central, square space with a dome, and two wings with freestanding, metal Corinthian columns that support the vault. At the end of each of the two wings stands a sculpture by Charles van der Stappen: *Dawn* and *Evening.* The trees alongside the passage and in the wings are cinnamon – identifiable from their decorative, pink and green leaves.

The luxuriantly flowering pink and purple clusters of flowers (*medinilla magnifica,* a tropical plant from the Philippines) in precious Chinese vases lend the Greenhouse a very refined atmosphere. The vases were brought back from the Far East in the 1860s by Leopold II – or the Duke of Brabant as he was

Charles van der Stappen, *Dawn,* late 19th century.

A corner of the estate with a gatehouse and a side-gallery of the Greenhouses.

at that time. The medinillas moved from their tropical hothouse to the Embarcadère a week before the Greenhouses were opened. Keeping the plants in the same vases year after year means that their roots have to be trimmed regularly – a far from easy task.

The adjoining compartment, which opens onto the estate, contains a copy of Donatello's *David.*

The visitor route now passes through part of the rolling and attractively laid out chateau grounds. To the left of the path is a knotty, strangely shaped *sophora japonica.* You now walk past the studio of Queen Elisabeth, the Palm Pond, the Iron Church and the Palm Pavilion to the Palm Greenhouse.

The Restoration of the Greenhouses

The Greenhouses have undergone major restoration work in recent decades. The warm and humid atmosphere inside and the rainy Belgian weather outside had an especially corrosive effect on the iron structure. The restoration work was led by a team of experts from the Belgian Buildings Authority – the government department responsible for maintaining state-owned buildings. Specialist firms developed modern techniques especially for these projects, the fabulous results of which now stand as a model for similar ventures elsewhere.

The repair work varied from light to extensive, depending on the degree of corrosion. The Theatre Greenhouse, for instance, had to be entirely dismantled and then reassembled with the necessary new components. The decorative ironwork was, however, recovered to as great an extent as possible, and the completed building matches Charles Girault's original design in every particular.

The 1980s saw the restoration of several buildings, including the Winter Garden, the Congo Greenhouse and the Palm Greenhouse. This was followed in the 1990s by the intensive restoration of the Orangery, the Banquet Greenhouse, the Theatre Greenhouse and the Maquet Greenhouse. Hortiplus glass has been used since 1981 because of its insulating and hence energy-saving properties. The slightly darker green paint that has been used on the iron and woodwork since 1992 closely matches the green used in the 19th century.

The restoration of the Maquet Greenhouse (or new Camellia Greenhouse) was completed in the spring of 2001, bringing the period of fundamental restoration to an end. All the same, the Greenhouses require constant care and systematic maintenance, and a number of structures will require extensive repair again within a few years, so that many more generations to come can enjoy the architectural beauty and botanical riches of this historical monument.

Sophora japonica by the path between the Embarcadère and the Palm Plateau.

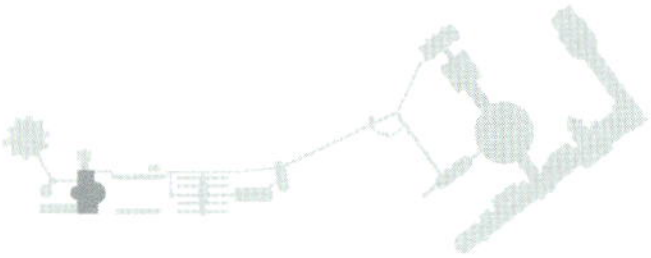

Geranium Gallery near the Palm Greenhouse.

A second group of greenhouses – the 'Palm Plateau' – is located some distance away from the Winter Garden complex. These structures were designed by Alphonse Balat and built in 1892–95. The most important metal constructions are the Palm Greenhouse and the Iron Church, which was altered in the 1930s.

From the grounds, you first enter a gallery with glass walls covered in geraniums, which you pass straight through into a courtyard. On the opposite side is the Débarcadère Greenhouse, which houses various types of bamboo. The Débarcadère opens onto Avenue du Parc Royal and was once the visitor's entrance. The glass barrel vault is supported by an elegant iron construction.

The atmospheric Loggia leads you to the Palm Greenhouse – a rectangular building with a high oval transept. The iron structural members are decorated with Art nouveau motifs.

A varied collection of tropical and subtropical trees and plants grows on either side of the tiled path. The ground is covered with green and yellow spikemoss and colourful pelargoniums. In the middle of the Greenhouse, on your right, you can see a giant palm tree in a pit. You will also recognise the

ladyfinger with its broad and very long leaves. The fruit of this tree – a small, hard banana – can only be eaten if it is boiled first. The gallery contains orchids.

A gallery on the left leads to the Iron Church (not open to the public). At the end of the Greenhouse, you walk through a group of giant rubber plants *(ficus elastica,* from India; hard, dark green, glossy leaves*)*. Nearby, on your right, you can see a white pisang (*strelitzia alba,* from South Africa). This plant has large, dark and torn leaves. Beyond the door is the staircase to the Palm Pavilion, which is also not open to the public. The stairs are bordered with hortensias. Another staircase on the right leads downwards via a small gallery. This takes you to the Azalea Greenhouse.

Alphonse Balat, Henri Maquet and Charles Girault

Plan of the Winter Garden and the Orangery showing the heating and ventilation systems, 1876.

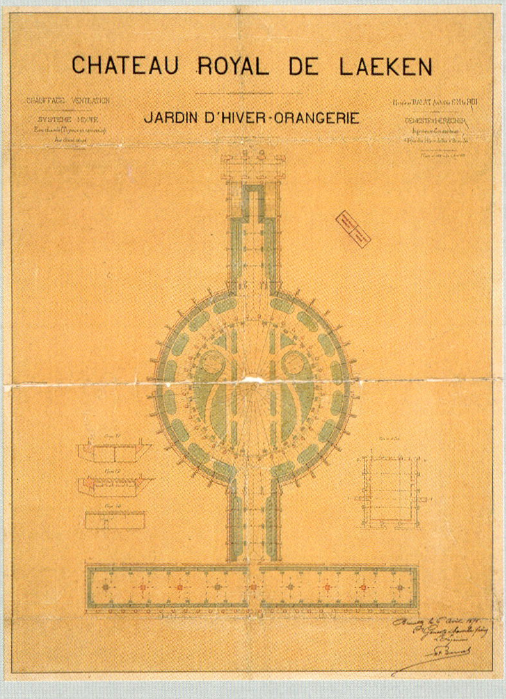

Between 1874 and 1905, Leopold II's favourite architects – Alphonse Balat, Henri Maquet and Charles Girault – all worked on the 'Glass Palace' at Laeken. The trio also designed several other major buildings commissioned by the King.

Alphonse Balat (1818–1895) was born in Gochenée, near Namur. Having studied at the Academy in Namur and later in Antwerp, he built and altered many chateaux and other residences for the aristocracy and haute bourgeoisie. He was appointed architect to the Duke of Brabant (the future King Leopold II) in 1860. For the remainder of his career he was primarily employed by Leopold – especially after the latter's coronation in 1865. He worked on the palaces at Brussels and Laeken, the chateaux at Ciergnon and Ardenne, and in Ostend. His designs included the Grand Staircase and Throne Room at the Royal Palace in Brussels, and he restored the Chateau at Laeken following the fire on 1 January 1890.

Balat was the most important architect to work on the Greenhouses at Laeken. He was a traditionalist, with a special love of classical styles and simplicity, who set out to achieve an effect of monumentality and harmony. All the same, he developed a new style of building in the Greenhouses, determined by their use of glass and iron. He died in Ixelles, shortly after the completion of the Iron Church.

Henri Maquet (1839–1909) trained at the Academy in Liège and then in Brussels, where he went on to design numerous houses. He was appointed 'architect-consultant' to the King in 1895 and built several greenhouses at Laeken and a gallery in Ostend on Leopold II's behalf. He also designed the new façade of the Royal Palace in Brussels, but – like the King himself – died before it was completed.

The French architect **Charles Girault** (1851–1932) was introduced to Leopold II at the 1900 Exposition Universelle in Paris, for which he had designed the Petit Palais. Girault became one of the King's favourite architects in the final decade of his reign. His principal works

Cross-section of the Congo Greenhouse, drawing by Alphonse Balat, 1886.

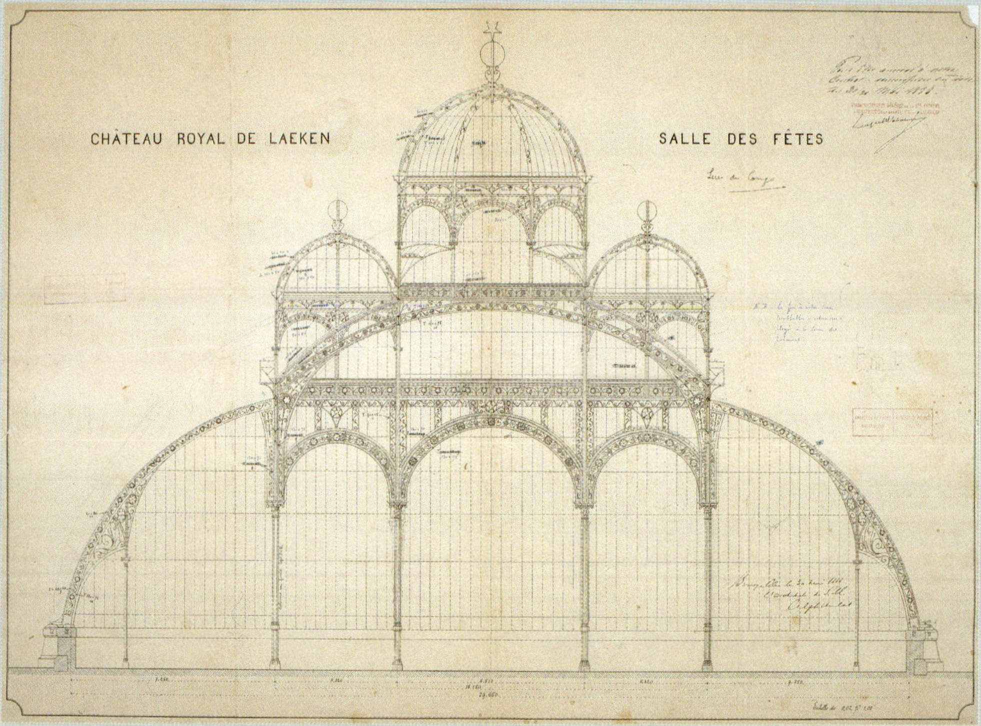

The restored Theatre Greenhouse.

The construction of the Winter Garden, 1874–1876.

Detail of the beams in the Winter Garden,
from A. Vierendeel, *La construction architecturale en fonte, fer et acier*, pl. 59, Paris-Brussels, 1897–1902.

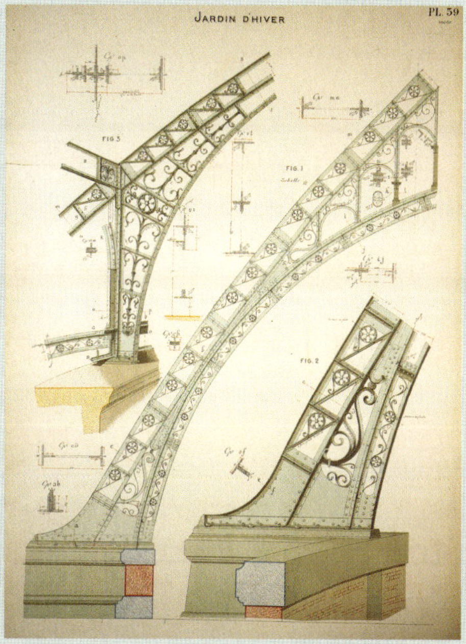

on the King's behalf were the Congo Museum in Tervuren, which was opened in 1910, and the Royal Galleries (a covered promenade) in Ostend, which he designed in 1906. He also designed two new wings for the Chateau at Laeken, including the Royal Stables and the Long Gallery with its Grand Staircase. The completion of that project was interrupted in 1909 by the death of the King. His successor, Albert I, had the work completed, but in a more sober spirit.

The plans that Girault produced for Leopold II were invariably ambitious projects – a World School in Tervuren, a grand plan for the development of the Belgian coast, a Palace of the Nation in Laeken with an underground rail link – only a small proportion of which ever came to fruition. Fabulous designs have been preserved, for instance, for highly impressive greenhouses. In reality, the Theatre Greenhouse was the only part of the Laeken complex designed by Girault.

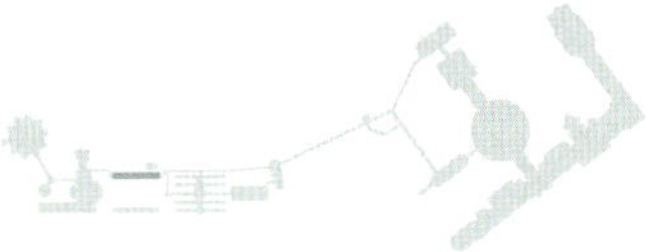

Leopold II had this simple hothouse built in 1885 especially for the cultivation of azaleas. For more than a century now, visitors to the Azalea Greenhouse have been treated to gorgeous arrangements with a wealth of varieties and colours.

The azalea, Flanders' best-known floral export, belongs to the genus *rhododendron*. The most common type at Laeken is *azalea indica* (originally from Japan). This is a typical hothouse azalea with countless varieties in terms of colour and type of flower. Several cultivars have been named after members of the Royal Family – Albert-Elisabeth (1921), Leopold-Astrid (1933), Fabiola (1978) and Mathilde (2001).

The plants are wintered in the Greenhouse at a temperature of 3° C. The interior is transformed in early April into a great show of floral decoration that is never exactly the same from one year to the next. Three gardeners work for two weeks to create the ensemble. It takes a special skill to ensure that all the different cultivars flower at the time of the opening.

A bust of Louis Paras, Head Gardener of the Royal Estate from 1921 to 1939, stands among the flowers. It was made by Queen Elisabeth in her studio.

Queen Elisabeth's Studio

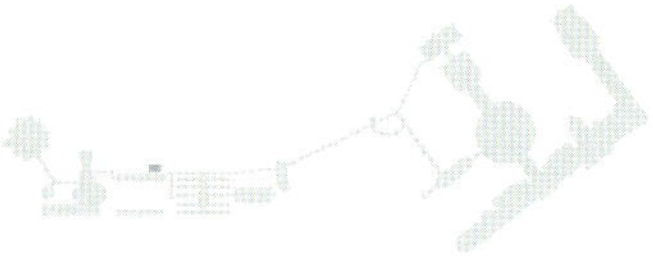

As you walked through the park, you passed by a picturesque building with a thatched roof. This is the sculpture studio of Queen Elisabeth (1876–1965) – wife of King Albert I (who reigned from 1909 to 1934). She is well known both for her humanitarian work during the First World War and for her immense interest in the arts. She did much to stimulate artistic life in Belgium and was herself a talented violinist and sculptor. The attractive studio, with its large window overlooking the estate, was built in 1938 and has been preserved in its original condition.

Queen Elisabeth lived in the Palm Pavilion for several years after the death of King Albert I. It is linked to the studio by the Palm Greenhouse and the Azalea Greenhouse.

The plant collection at the Royal Greenhouses in Laeken includes thousands of trees, shrubberies, flowers and plants. No fewer than 5,000 flowering plants are cultivated to decorate the Greenhouses in April and May, when they are open to the public.

The overall Greenhouse complex covers 1.5 hectares. The total surface area of the glass roofs is 2.5 hectares.

The distance between the Chateau (Theatre Gallery) to the Iron Church is 700 m, beneath a single, continuous glass roof – a unique concept in the history of greenhouse architecture.

The Orangery is 97 m long, 13 m wide and 8 m high.

The Winter Garden, with its two connecting greenhouses, covers 2,400 m^2 and is covered with 5,090 m^2 of glass. The Congo Greenhouse is 930 m^2 in size (with 1,518 m^2 of glass) and the Palm Greenhouse 604 m^2 (with 1,308 m^2 of glass).

The Winter Garden is 25.648 m high. Its large dome is supported by a colonnade of 36 pillars on which rest 36 iron beams. The total interior diameter of this greenhouse, including the gallery, is 57.124 m. The gallery is 7.935 m across. The dome has a diameter of 41.254 m and is surmounted with a twelve-sided corolla with a diameter of 8 m. Each of the twelve petals in the lantern supports the pressure of three beams.

The construction of the Winter Garden (1874–'76) required 651.534 tons of iron.

The basic structure of the Congo Greenhouse is a square with 30 m sides. Four iron arches measuring 15 m, each take in three arcades in the middle of the Greenhouse. This central section is surmounted by an octagonal dome and surrounded by four small square domes.

Exotic trees and plants require warmth and humidity. The Greenhouse complex is heated with 14 boilers (overall capacity: 8 million Kcal/h). The oil tank can hold a total of 200,000 litres. The average consumption is around 800,000 litres of oil a year.

It takes 200,000 litres of hot water to heat the entire complex, a quarter of which (56,902 litres) goes to the Winter Garden. The hot water is carried by a system of pipes running for a total of 14 km – the channels with the hot-water pipes are covered with wrought-iron grates.

Water from a large pond on the estate – the Carp Pond – is used to water the plants and to maintain humidity. This is done using a pumping station with two pumps operating at a rate of 10,000 litres per hour. There are three distribution stations in the basement of the Greenhouses. The maximum flow-rate of 52,000 litres per day is used on hot days – especially in the Winter Garden, the Congo Greenhouse and the Palm Greenhouse.

The roofs are made using a special type of glass – Hortiplus, which is 4 mm thick. The outside of the glass is covered with a thin layer of stable tin oxide, giving it an extra insulating effect. It also creates attractive colour effects.

There are 44 orange trees, most of which are over 200 years old and belonged to the original 18th-century collection.

Some of the palm trees are more than a hundred years old and were acquired by Leopold II himself. The oldest palms are believed to be about 150 years old.

The 296 different varieties of camellia form the oldest and most important collection of its kind under glass in the world. In addition to their historical and cultural importance, they are extremely important in scientific terms because of the genetic material they contain.

The Congo Greenhouse.

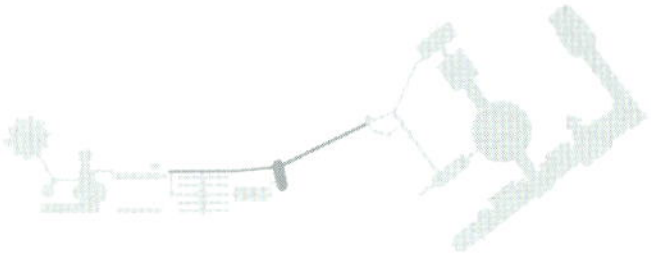

Balat linked the Palm Plateau with the Winter Garden complex via the Long Gallery – a glass passageway running for 200 m. The first section – the Geranium Gallery – runs above ground to the Mirror Greenhouse, while the second – the Underground Gallery – leads to the Embarcadère.

Tall geraniums in a variety of colours grow along the walls of the Geranium Gallery. Masses of colourful fuchsias hang down from above, interspersed with the sweet-smelling purple and white flowers of the heliotrope and the dark brown, yellow and red Chinese lanterns of the abutilon, which recalls Belgium's red, yellow and black tricolour. Another striking exhibit is the bush *(plumbago)* with the small, bright blue flowers at the entrance to the Long Gallery.

The geraniums *(pelargonium)* are not the climbing type, but vigorous growers that can reach a height of two metres. These too were cultivated intensively in the Greenhouses at Laeken in Leopold II's time. There were no fewer than 360 varieties here at the beginning of the 20th century.

The Diana Greenhouse.

The Fern Cross, with its eight hothouses that are not open to the public, is located between the Azalea Greenhouse and the Diana Greenhouse, and runs parallel to the Geranium Gallery. Several of the side-galleries do, however, offer a charming glimpse. One wing of the Fern Cross is laid out as an idyllic resting spot.

Balat built a small pavilion – the Diana Greenhouse – in a bend halfway between the Azalea Greenhouse and the Mirror Greenhouse. A statue of Diana, the divine huntress, stands towards the rear. The plants in this greenhouse are especially attractive, including as they do, orchids, palms, a tall tree fern (*cyathea* or *alsophila australis,* from Australia and Tahiti; rough trunk, hairy stalks and fine leaves with spurs) and a fine evergreen dombeya bush (*astrapaea,* from Madagascar; large, matt green leaves with serrated edges and, in February–March, clusters of brilliant scarlet flowers).

The Geranium Gallery.

A greenhouse is an artificial environment in which people optimise plant growth by controlling the climate. This requires a thorough knowledge of the natural surroundings in which the various plants originated.

The most important types of greenhouses in Laeken are tropical, subtropical and cool. The tropical greenhouses are always hot (temperature above 18 ° C) and humid. The subtropical ones – such as the Winter Garden, Congo Greenhouse and Palm Greenhouse – are a little less warm (above 10 ° C) but still humid. The temperature in the cool greenhouses is 3 ° C.

The climate within the greenhouse is determined by its particular combination of temperature, humidity and light. The right temperature is obtained in winter by heating and in the summer through ventilation. The correct humidity level is obtained by watering and spraying. The incoming light is toned down between March and September by whitewashing the glass or

The Underground Gallery.

covering it with tarpaulins. During the gloomy winter months, by contrast, there is insufficient light for the tropical and subtropical plants, and so temperature is adjusted according to the light.

There are two main aspects to the day-to-day management of the Royal Estate. The first consists of cultivating and caring for the collection of plants, and the second of restoring and maintaining the greenhouse structure and the technical systems on which it relies. Tending for the plants includes such elements as climate control (temperature, humidity and light), pruning, removing dead growth, applying fertiliser, renewing the soil, planting, potting, and preventing and treating disease, mould and insects.

While the unseen elements – the heating, electricity and water systems – have been entirely modernised, the Greenhouse buildings and the plant collection still match Leopold II's original conception – a testimony to the intensive care lavished on them.

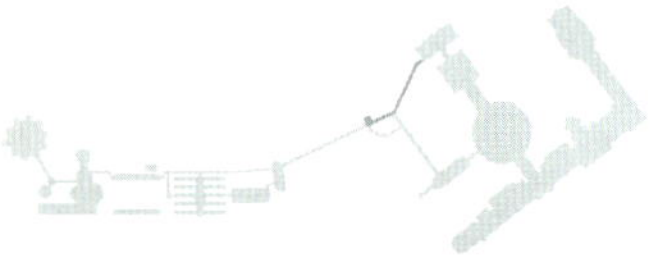

The Geranium Gallery opens into the Mirror Greenhouse – the smallest of the greenhouses in this glass city. It is located at the foot of the staircase leading from the section above ground to the Underground Gallery. A cinnamon tree grows in a corner of the staircase, while a large tree fern stands at the bottom, next to the banister, which is festooned with *fatshedera lizei* (a cross between ivy and *fatsia japonica*). The most striking feature, however, is the imposing royal fern that hangs roughly in the middle of the Greenhouse. The mirror reflects a mixture of flowers and plants, most notably a large, silvery green staghorn fern (*platycerium bifurcatum,* from Australia) above the passageway leading to the Underground Gallery.

The gallery's walls are covered with a severely pruned climbing fig tree (*ficus pumila,* from China and Japan) and oak ferns (*polypodium*), interspersed with a whole collection of staghorn ferns. The oak ferns have attached themselves to the hundred-year-old climbing fig. Beneath them stand snail-shaped vases in amongst maidenhair ferns (*adiantum*).

You now find yourself back at the Embarcadère. Walking through the exotic iron and glass paradises of the Winter Garden complex will take you to the Orangery.

The Greenhouse complex at Laeken comprises three zones. The first is formed by the line of monumental buildings consisting of the Embarcadère, Congo Greenhouse, Winter Garden and Orangery, which is linked to the Chateau via the Theatre Greenhouse. The axis connecting the Embarcadère and the Orangery is 150 m long. The second zone is the 'Palm Plateau' with the Débarcadère, Palm Greenhouse and Palm Pavilion, Iron Church, Azalea Greenhouse and the hothouses. The third zone consists of the series of above-ground and underground galleries and glass passageways with their small but atmospheric linking pavilions that connect the two large groups of greenhouses.

The first and oldest zone was conceived by Leopold II as a single large reception space. You cannot help but be impressed by the powerful display when you take the stairs from the Embarcadère up to the Congo Greenhouse and then enter the Winter Garden. An unreal world of iron, glass, subtropical plants, light and shade unfolds enchantingly before your eyes.

We offer our respectful gratitude to
Their Majesties the King and Queen
and thank their staff
for the willing assistance
they provided during
the compilation of this guide.

Bibliography

The following books are recommended to those who would like to know more about the history and architecture of the Royal Greenhouses at Laeken, and the wealth of plants and flowers they contain:

L. Ranieri, *Léopold II, urbaniste,* Hayez, Brussels 1973.

R. Vandendaele et al., *Poelaert et son temps,* exhibition catalogue (Institut Supérieur d'Architecture Victor Horta), City of Brussels/Crédit Communal de Belgique, Brussels 1980.

B. Emerson, *Léopold II, le royaume et l'empire,* Paris/Gembloux 1980.

E. Vandewoude, J. Vandenbreeden and P. Van Gorp, *The Royal Glasshouses at Laeken: Eight Plates by Margot Weemaes: with an Historical, Architectural Botanical Introduction,* Koninklijke Schenking, Brussels 1981. This publication was an important source for the compilation of this LUDION Guide.

E. Vandewoude, 'De bouw van de Wintertuin te Laken, 1874–1876. Een initiatief van Leopold II', *Album Carlos Wyffels,* Brussels 1987, pp. 457–473.

H. van den Bossche, 'Koninklijke camellias uit het land van de rijzende zon?', *Monumenten en Landschappen,* Brussels 1989, pp. 41–58.

H. Balthazar and J. Stengers (eds), *Dynastie et culture en Belgique,* Fonds Mercator, Antwerp 1990.

P. Lombaerde and R. Gobyn, *Léopold II Roi-Batisseur,* Pandora, Ghent 1995.

E. Goedleven, *The Royal Greenhouses of Laeken,* Lannoo/SFI, Brussels 1997. This attractively illustrated book provides a fairly comprehensive survey of the Royal Greenhouses' construction, architecture and plant collection. Together with the personal notes of the director of the *Domaine royal de Laeken,* it was a crucial source of information during the preparation of this LUDION Guide, especially for the botanical descriptions and the review of the plants in the different greenhouses.

S. Basch, *Léopold de Belgique (futur Léopold II). Voyage à Constantinople, 1860,* Complexe, Brussels 1997.

G. Janssens and J. Stengers, *Nouveaux regards sur Léopold I & Léopold II: fonds d'archives Goffinet,* Fondation Roi Baudouin, Brussels 1997.

A. Deseyne, *Elisabeth. Beeldhouwwerken,* exhibition catalogue, Bruges 1997.

Les serres royales de Laeken, Service de presse de la Régie des bâtiments, Brussels 1998.

INDEX OF TREES AND PLANTS

Camellia teutonia, from A. and A. Verschaffelt, *Monographie des camélias*, Ghent, 1860.

Photography
Ch. Bastin & J. Evrard: pp. 2, 6, 11 (right), 19, 20 (upper left, upper right), 22, 23, 27, 28, 31, 54, 55, 60, 62, 64
Service fédéral d'information: pp. 14, 15, 29, 30, 32, 33, 34, 46, 47, 51, 61
Johan Jacobs: back cover
Paul Maeyaert: pp. 4, 8, 9, 11 (left), 12, 13, 16, 18, 20 (bottom), 21, 26, 35, 36–37, 39, 41, 44, 48, 49, 52–53, 56, 57, 58, 59
Oswald Pauwels (Monumenten & Landschappen): 17, 63
PhotoNews: p. 10
Luc Schrobiltgen: pp. 42, 43

p. 24: A. Vierendeel, *La construction architecturale en fonte, fer et acier,* Paris-Brussels, 1897–1902, II 71.933 C in f°
p. 45: Archives of the Royal Palace

Every effort has been made to trace copyright holders. If, however, you feel that you have inadvertently been overlooked, please contact the publisher: Ludion, Muinkkaai 42, B-9000 Ghent; Herengracht 376, NL-1016 CH Amsterdam.

All rights reserved. No part of this publication may be reproduced or transmitted, in any form or by any means, electronic or mechanical, including photocopy, recording or any other information storage and retrieval system without prior permission in writing from the publisher.

© 2001 Ludion Ghent-Amsterdam & Irene Smets
© 2001 SOFAM
Design: Antoon De Vylder, Herentals
Typesetting: De Diamant Pers, Herentals
Translation: Ted Alkins, Heverlee
Colour separations and printing: Die Keure, Bruges
D/2001/6328/10
ISBN: 90-5544-339-5